MW00836917

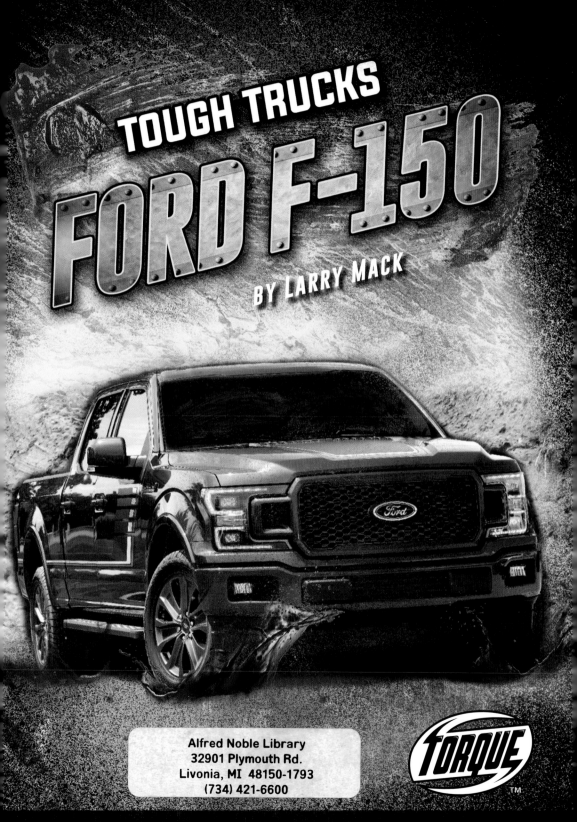

TOUGH TRUCKS

FORD F-150

BY LARRY MACK

BELLWETHER MEDIA • MINNEAPOLIS, MN

Are you ready to take it to the extreme?
Torque books thrust you into the action-packed world
of sports, vehicles, mystery, and adventure. These books may
include dirt, smoke, fire, and dangerous stunts.
WARNING read at your own risk.

This edition first published in 2019 by Bellwether Media, Inc.

No part of this publication may be reproduced in whole or in part without written permission of the publisher.
For information regarding permission, write to Bellwether Media, Inc., Attention: Permissions Department,
6012 Blue Circle Drive, Minnetonka, MN 55343.

Library of Congress Cataloging-in-Publication Data

Names: Mack, Larry, author.
Title: Ford F-150 / by Larry Mack.
Description: Minneapolis, MN : Bellwether Media, Inc., 2019. | Series:
   Torque: Tough Trucks | Includes bibliographical references and index. |
Audience: Ages 7-12.
Identifiers: LCCN 2018002185 (print) | LCCN 2018006819 (ebook) | ISBN
   9781626178922 (hardcover : alk. paper)| ISBN 9781681036113 (ebook)
Subjects: LCSH: Ford F-Series trucks–Juvenile literature.
Classification: LCC TL230.5.F57 (ebook) | LCC TL230.5.F57 M32 2019 (print) |
   DDC 629.223/2–dc23
LC record available at https://lccn.loc.gov/2018002185

Editor: Betsy Rathburn   Designer: Josh Brink

TABLE OF CONTENTS

AT THE RACES

The racers drive day and night. They do not sleep for nearly a day and a half!

This is the Baja 1000. It is the world's most famous **off-road** race. Cars, trucks, and motorcycles speed across Mexico's desert. They charge over mountain roads!

The 2017 race is about 850 miles (1,368 kilometers) long. Teams must complete the race in less than 36 hours.

DOUBLE DUTY

AFTER THE BAJA 1000, FOUTZ DROVE HIS FORD 400 MILES (644 KILOMETERS) HOME FROM THE RACE!

Racer Greg Foutz and his co-driver blast across the finish line with seconds to spare! They have completed one of the world's toughest races. And they did it with a Ford F-150 Raptor!

7

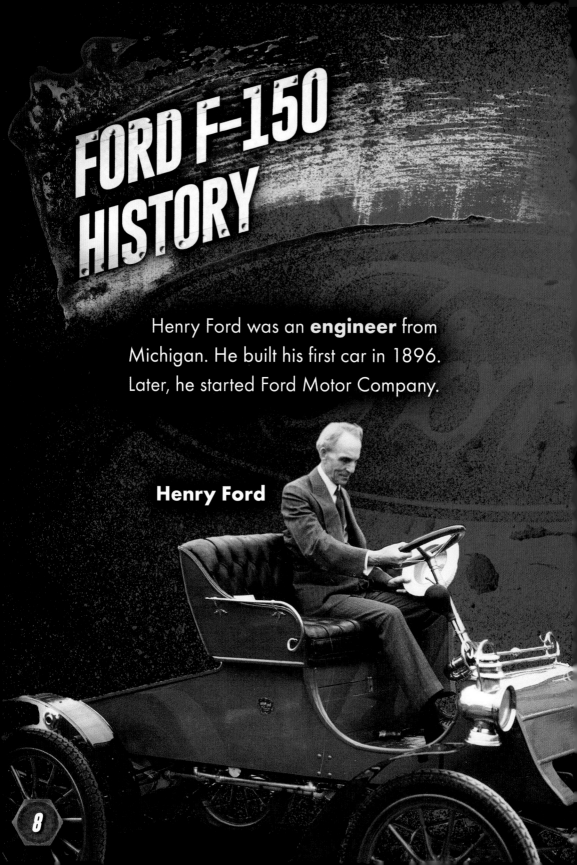

FORD F-150 HISTORY

Henry Ford was an **engineer** from Michigan. He built his first car in 1896. Later, he started Ford Motor Company.

Henry Ford

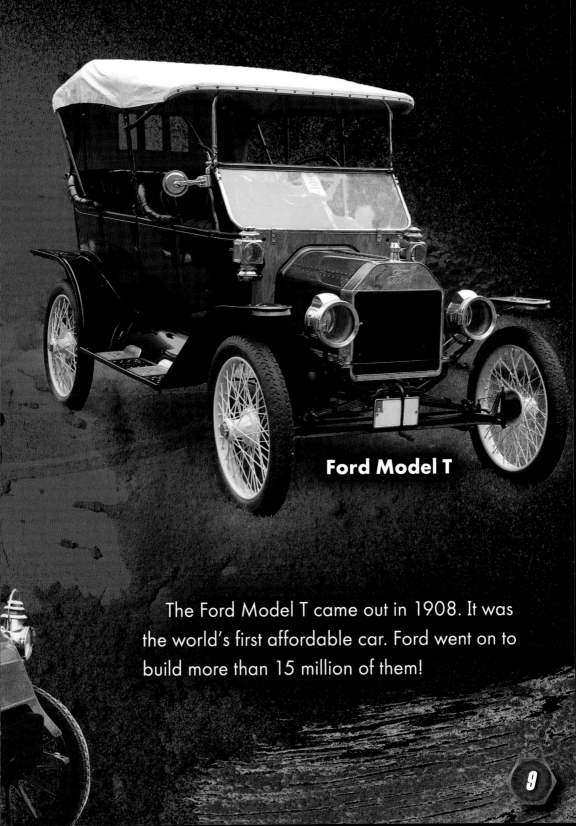

Ford Model T

The Ford Model T came out in 1908. It was the world's first affordable car. Ford went on to build more than 15 million of them!

The company began making truck **chassis** in 1917. Buyers added their own truck bodies to the frame. In 1925, Ford began selling complete trucks.

1925 Ford Model T

1976 Ford F-150

Ford F-Series trucks came out in 1948. They were comfortable and easy to drive. In 1975, Ford added the F-150 to the F-Series. It would become one of the most famous pickups ever!

FORD F-150 TODAY

Today, Ford F-150s have powerful engines. Light **alloy** bodies help the trucks use less fuel. Ford F-150s also have cool gadgets to improve safety.

Ford offers many **models** of the F-150. Each has different features and uses. For example, some King Ranch F-150s can haul more than 7,000 pounds (3,175 kilograms)!

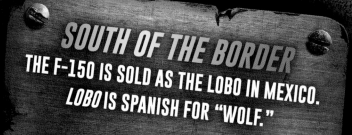

SOUTH OF THE BORDER
THE F-150 IS SOLD AS THE LOBO IN MEXICO. LOBO IS SPANISH FOR "WOLF."

Ford F-150 King Ranch

FEATURES AND TECHNOLOGY

Pickups need a lot of power to do their jobs. Some F-150s come with an EcoBoost engine. This powerful engine is available with six or eight **cylinders**. The engine's **turbocharger** forces air into the cylinders. There, it mixes with fuel. This gives the EcoBoost extra power!

FUEL ECONOMY

FORD ECOBOOST ENGINES MAY HAVE SAVED DRIVERS 110 MILLION GALLONS (416 MILLION LITERS) OF FUEL IN 2017.

3.5L V6 EcoBoost engine

Ford F-150 XLT

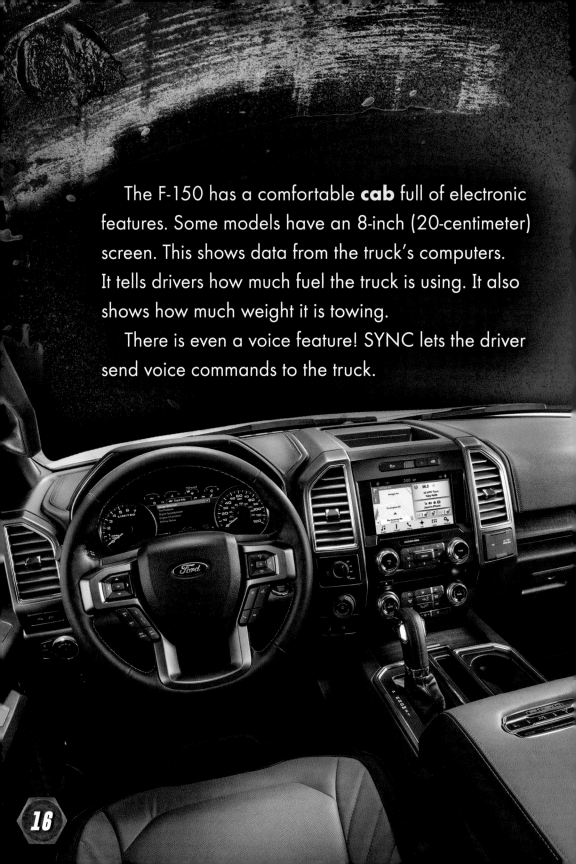

The F-150 has a comfortable **cab** full of electronic features. Some models have an 8-inch (20-centimeter) screen. This shows data from the truck's computers. It tells drivers how much fuel the truck is using. It also shows how much weight it is towing.

There is even a voice feature! SYNC lets the driver send voice commands to the truck.

2018 FORD F-150 RAPTOR SPECIFICATIONS

ENGINE	3.5L V6 ECOBOOST ENGINE
HORSEPOWER	450 HP (336 KILOWATTS) @ 5,000 RPM
TORQUE	510 LB-FT (71 KG-M) @ 3,500 RPM
TOWING CAPACITY	UP TO 8,000 POUNDS (3,629 KILOGRAMS)
MAXIMUM PAYLOAD	1,200 POUNDS (544 KILOGRAMS)
FUEL ECONOMY	15 TO 18 MILES PER GALLON
CURB WEIGHT	UP TO 5,800 POUNDS (2,631 KILOGRAMS)
WHEEL SIZE	17 INCHES (43 CENTIMETERS)

Some features keep drivers safe. Lights in the side mirrors warn drivers to stay in their lane. They blink when another vehicle is passing.

Some F-150 models have four cameras. They send images from all around the truck to the screen in the cab. The images show the driver things they cannot see outside the truck!

AT THE TRACK
FORD F-150S ARE RACED IN THE NASCAR TRUCK SERIES.

cab

TRUCK OF THE FUTURE

The Ford F-150 has changed a lot in the past 40 years. But the truck is as popular as ever! Many drivers choose this comfortable and safe pickup.

Ford will continue to update the F-150 with new technology. From races to work sites, pickup fans love the F-150. No wonder it is the world's most popular pickup!

HOW TO SPOT A FORD F-150

RECTANGULAR GRILLE

SMOOTH LINES ON HOOD

C-SHAPED HEADLIGHTS

GLOSSARY

alloy—a combination of aluminum and other metals; some alloys are stronger than steel.

cab—the area of a pickup where the driver and passengers sit

chassis—the frames of pickup trucks

cylinders—chambers in an engine in which fuel is ignited

engineer—a person who designs and builds cars and other machines

models—specific kinds of trucks

off-road—taking place on unpaved roads

turbocharger—a device that forces air into an engine's cylinders to help create power

TO LEARN MORE

AT THE LIBRARY

Bowman, Chris. *Pickup Trucks*. Minneapolis, Minn.: Bellwether Media, 2018.

Mack, Larry. *Chevrolet Silverado*. Minneapolis, Minn.: Bellwether Media, 2019.

Mack, Larry. *Ram 1500*. Minneapolis, Minn.: Bellwether Media, 2019.

ON THE WEB

Learning more about the Ford F-150 is as easy as 1, 2, 3.

1. Go to www.factsurfer.com.

2. Enter "Ford F-150" into the search box.

3. Click the "Surf" button and you will see a list of related web sites.

With factsurfer.com, finding more information is just a click away.

INDEX

The images in this book are reproduced through the courtesy of: Ford, front cover (hero), pp. 4-5, 6-7, 12-13, 13, 14-15, 15-16, 16, 18-19, 20-21, 21 (left, middle, right); Andrey Kuzmin, front cover (title texture), pp. 7 (metal), 12 (metal), 14 (metal), 18 (metal); marinya, front cover (background); xpixe, front cover (top mud, bottom mud); diogoppr, front cover (mud splash); Roberto Lusso, pp. 2-3 (logo); Steve Lagreca, pp. 2-3 (truck); Yulia Plekhanova, pp. 2-3 (background); SZ Photo/ Scherl/ Alamy, pp. 8-9; Lothar Spurzem/ Wikipedia, p. 9; Sicnag/ Wikipedia, p. 10; Greg Gjerdingen/ Flickr, p. 11; Ed Aldridge, p. 14; Nigel Kinrade/ AP Images, p. 17; sociologas, p. 21 (metal).